J
566
W

Wexo, John Bonnett.
Flyers / by John Bonnett Wexo ; chief artist, Walter Stuart. -- Mankato, MN : Creative Education, Inc., 1991, c1989.

24 p. : ill. -- (Prehistoric zoobooks ; book 8)

51160

SUMMARY: Discusses the three groups of prehistoric vertebrates that took to the sky, the flying reptiles, the first birds, and the first bats.
ISBN 0-88682-394-3(lib.bdg.) : $11.90

1. Pterosaurs. 2. Birds, Fossil. 3. Bats, Fossil. 4. Prehistoric animals. I. Title. II. Series.

14

90-46216 /AC
MARC

FLYERS

On The Cover:
*A Pterodactyl **(TER-uh-DAK-till)**.*
Reptiles were the first vertebrates
that had wings and could fly.
Cover Art by Walter Stuart.

Published by Creative Education, Inc., 123 South Broad Street, Mankato, Minnesota 56001

Printed by permission of Wildlife Education, Ltd.

ISBN 0-88682-394-3

Created and written by
John Bonnett Wexo

Chief Artist
Walter Stuart

Senior Art Consultant
Mark Hallett

Design Consultant
Eldon Paul Slick

Production Art Director
Maurene Mongan

Production Artists
Bob Meyer
Fiona King
Hildago Ruiz

Photo Staff
Renee C. Burch
Katharine Boskoff

Publisher
Kenneth Kitson

Associate Publisher
Ray W. Ehlers

FLYERS

This Volume is Dedicated to: Kenneth Kitson, Ray Ehlers, Gerald Marino and all of my other friends at Wildlife Education. Their enduring support and patience have sustained me on the long road to completion of these books.

Art Credits

Page Eight: Timothy Hayward; **Page Nine: Upper Left and Lower Right,** Walter Stuart; **Upper Right,** Timothy Hayward; **Page Ten: Middle Left,** Walter Stuart; **Lower Left,** Timothy Hayward; **Pages Ten and Eleven:** Timothy Hayward; **Page Eleven: Upper Right,** Walter Stuart; **Middle Left and Right,** Robert Bampton; **Lower Right,** Walter Stuart; **Page Twelve: Lower Left and Upper Right,** Timothy Hayward; **Pages Twelve and Thirteen:** Timothy Hayward; **Page Thirteen: Upper Middle,** Walter Stuart; **Upper and Lower Middle:** Robert Bampton; **Lower Left,** Walter Stuart; **Lower Middle and Lower Right,** Robert Bampton; **Page Fourteen: Middle Left and Lower Right,** Robert Bampton; **Lower Left,** Timothy Hayward; **Pages Fourteen and Fifteen: Center,** Timothy Hayward; **Page Fifteen: Upper Left,** Robert Bampton; **Upper Right,** Timothy Hayward; **Middle and Middle Right,** Walter Stuart; **Lower Right,** Walter Stuart; **Page Sixteen:** Timothy Hayward; **Page Seventeen: Top, Middle Right and Lower Left,** Robert Bampton; **Middle Left and Lower Right,** Timothy Hayward; **Page Eighteen: Lower Right,** Robert Bampton; **Pages Eighteen and Nineteen:** Timothy Hayward; **Page Nineteen: Middle Left,** Robert Bampton; **Page Twenty: Middle Left,** Robert Bampton; **Pages Twenty and Twenty-one:** Timothy Hayward; **Pages Twenty-two and Twenty-three: Background,** Timothy Hayward; **Figures,** Chuck Byron.

Photographic Credits

Pages Six and Seven: Gordon Menzie *(Model by Andrea von Sholly);* **Background,** Brian Brake, Nat'l Audubon Society Collection *(Photo Researchers);* **Page Sixteen: Left,** Gordon Menzie *(Model by Walter Stuart);* **Lower Right,** John Ostrom; **Page Twenty: Upper Right,** L. Riley *(Bruce Coleman, Inc.);* **Middle Left,** A.N.T. *(NHPA);* **Lower Left,** Gordon Menzie *(Model by Walter Stuart).*

Creative Education would like to thank Wildlife Education, Ltd., for granting them the right to print and distribute this hardbound edition.

Contents

There were very good reasons for leaving the ground . . .

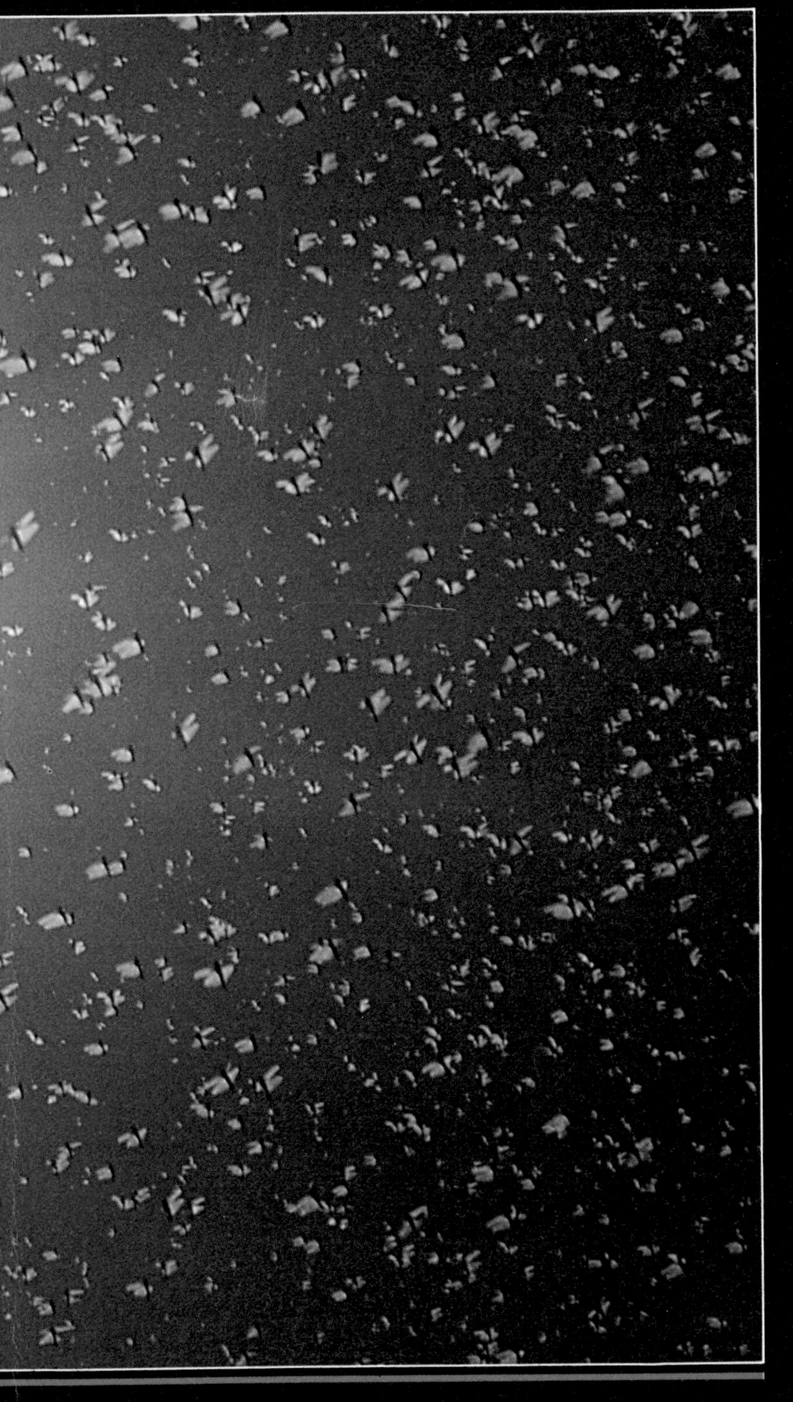

BOOK EIGHT

Flyers

During the Age of Dinosaurs, the dinosaurs truly "ruled the land." But some other groups of vertebrates found a way to live in a world *that dinosaurs could not reach*—the world of **the air**. Three separate groups of vertebrates **started to fly**—the birds, the flying reptiles, and the bats. As you will see, it was not easy for vertebrates to get up into the air. But there were **great rewards** for animals that could fly. They could reach new sources of **food**, and they would be **much safer** up in the air . . .

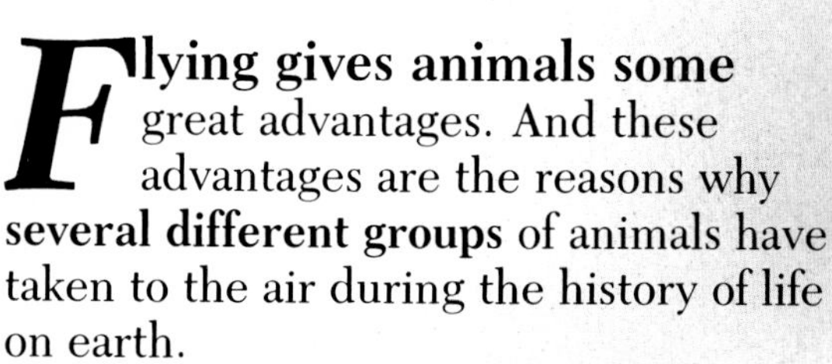

Flying gives animals some great advantages. And these advantages are the reasons why **several different groups** of animals have taken to the air during the history of life on earth.

As you remember, **insects** were the first animals to fly. Some insects developed wings to escape from predators—and to find new sources of food. For the same reasons, some **reptiles** started to fly, and **birds** and **bats** followed them into the air.

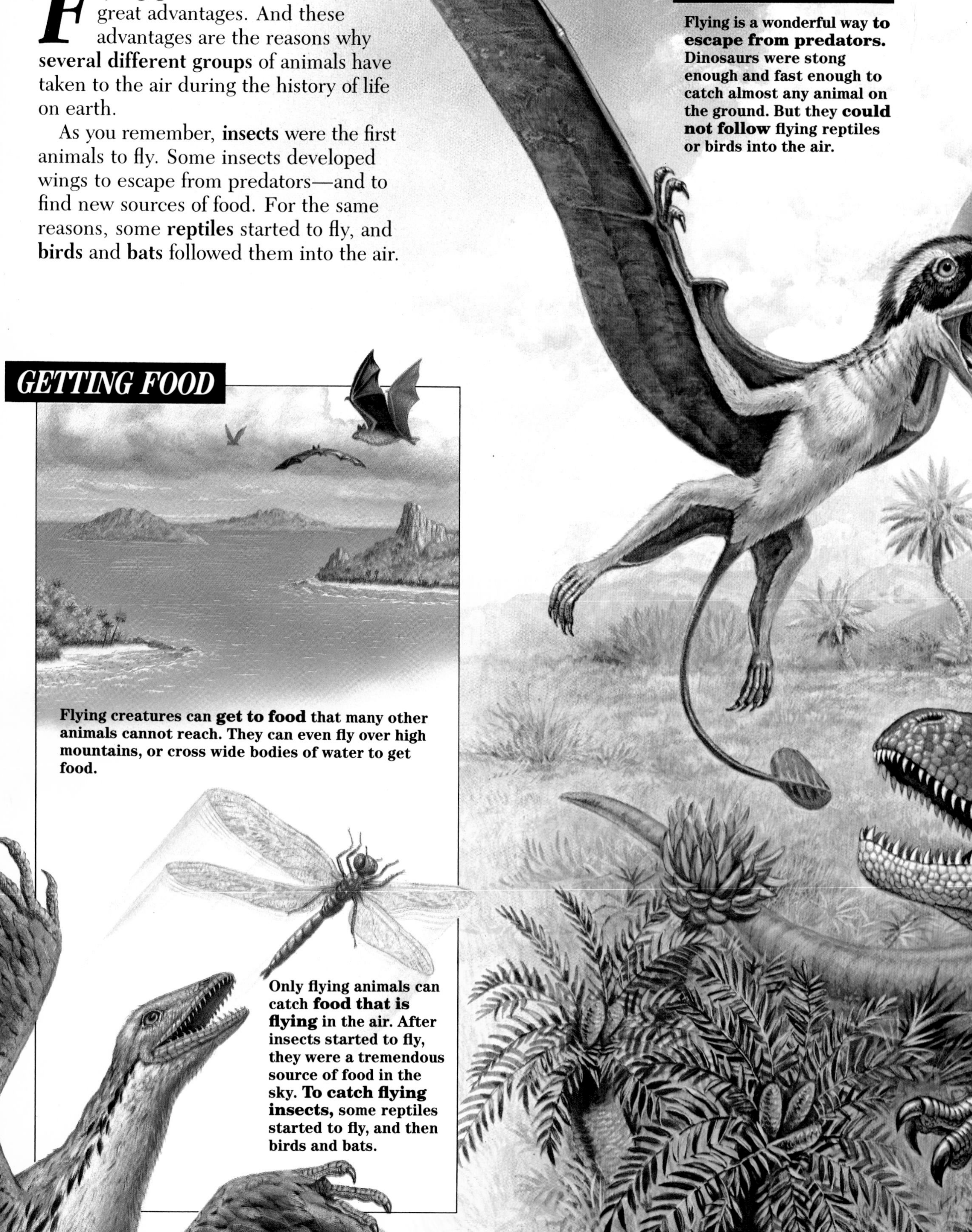

GETTING AWAY

Flying is a wonderful way **to escape from predators.** Dinosaurs were stong enough and fast enough to catch almost any animal on the ground. But they **could not follow** flying reptiles or birds into the air.

GETTING FOOD

Flying creatures can **get to food** that many other animals cannot reach. They can even fly over high mountains, or cross wide bodies of water to get food.

Only flying animals can catch **food that is flying** in the air. After insects started to fly, they were a tremendous source of food in the sky. **To catch flying insects,** some reptiles started to fly, and then birds and bats.

STAYING SAFE

Up in the trees, or high on a rocky cliff, flying animals were safe from many predators. They could build nests and **raise their young** in greater safety.

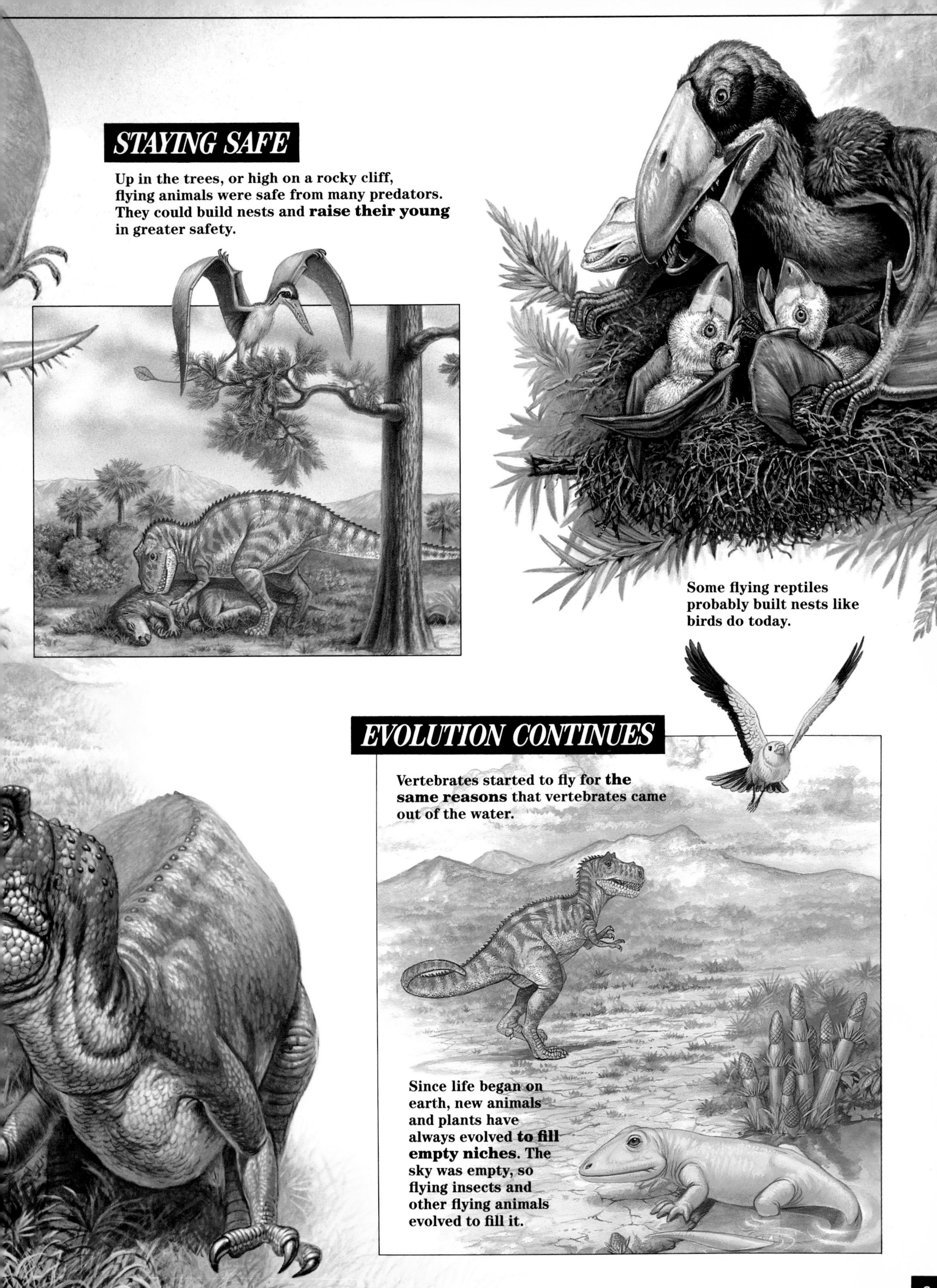

Some flying reptiles probably built nests like birds do today.

EVOLUTION CONTINUES

Vertebrates started to fly for **the same reasons** that vertebrates came out of the water.

Since life began on earth, new animals and plants have always evolved **to fill empty niches.** The sky was empty, so flying insects and other flying animals evolved to fill it.

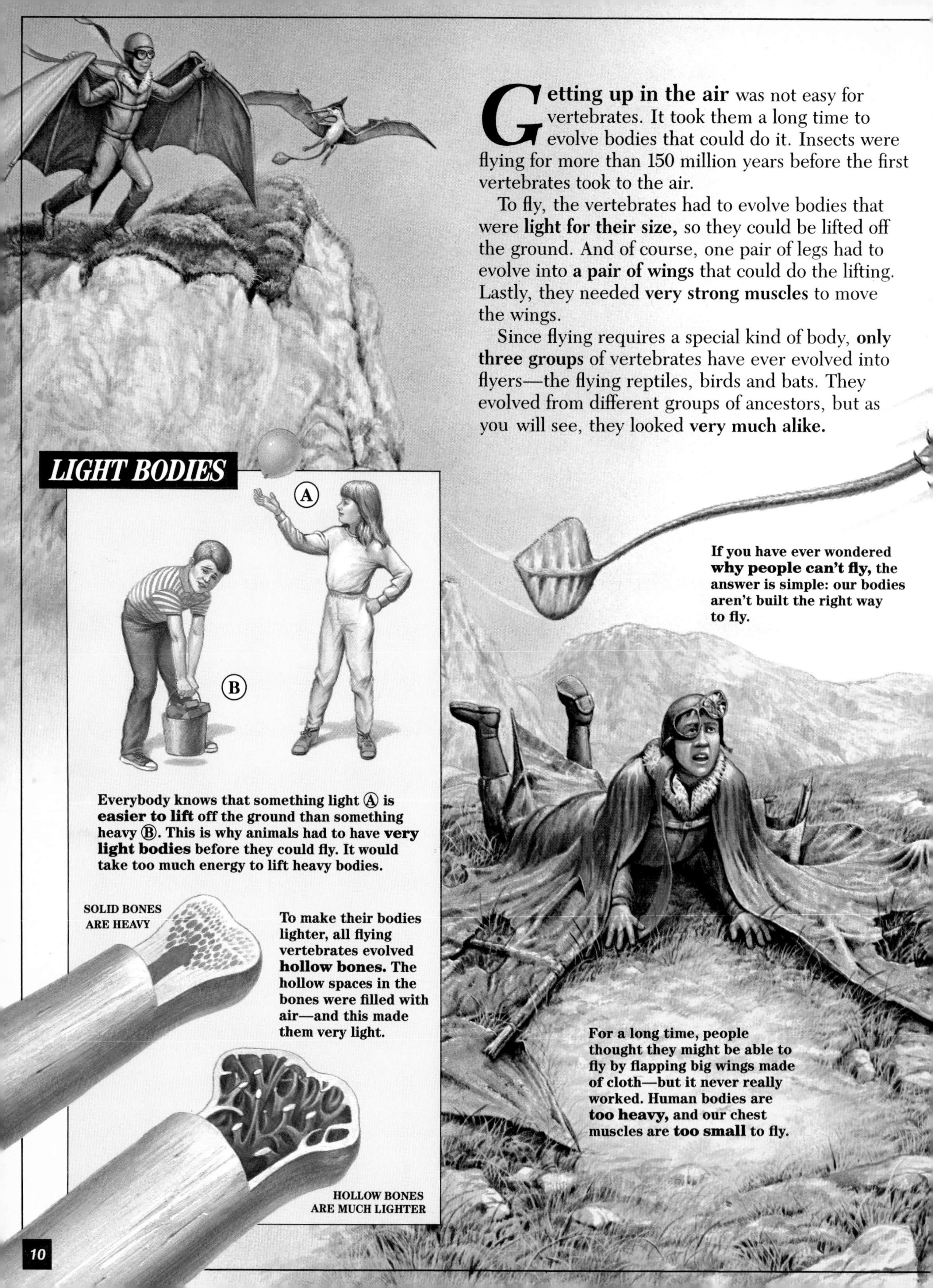

Getting up in the air was not easy for vertebrates. It took them a long time to evolve bodies that could do it. Insects were flying for more than 150 million years before the first vertebrates took to the air.

To fly, the vertebrates had to evolve bodies that were **light for their size,** so they could be lifted off the ground. And of course, one pair of legs had to evolve into **a pair of wings** that could do the lifting. Lastly, they needed **very strong muscles** to move the wings.

Since flying requires a special kind of body, **only three groups** of vertebrates have ever evolved into flyers—the flying reptiles, birds and bats. They evolved from different groups of ancestors, but as you will see, they looked **very much alike.**

LIGHT BODIES

Everybody knows that something light Ⓐ is **easier to lift** off the ground than something heavy Ⓑ. This is why animals had to have **very light bodies** before they could fly. It would take too much energy to lift heavy bodies.

To make their bodies lighter, all flying vertebrates evolved **hollow bones.** The hollow spaces in the bones were filled with air—and this made them very light.

If you have ever wondered **why people can't fly,** the answer is simple: our bodies aren't built the right way to fly.

For a long time, people thought they might be able to fly by flapping big wings made of cloth—but it never really worked. Human bodies are **too heavy,** and our chest muscles are **too small** to fly.

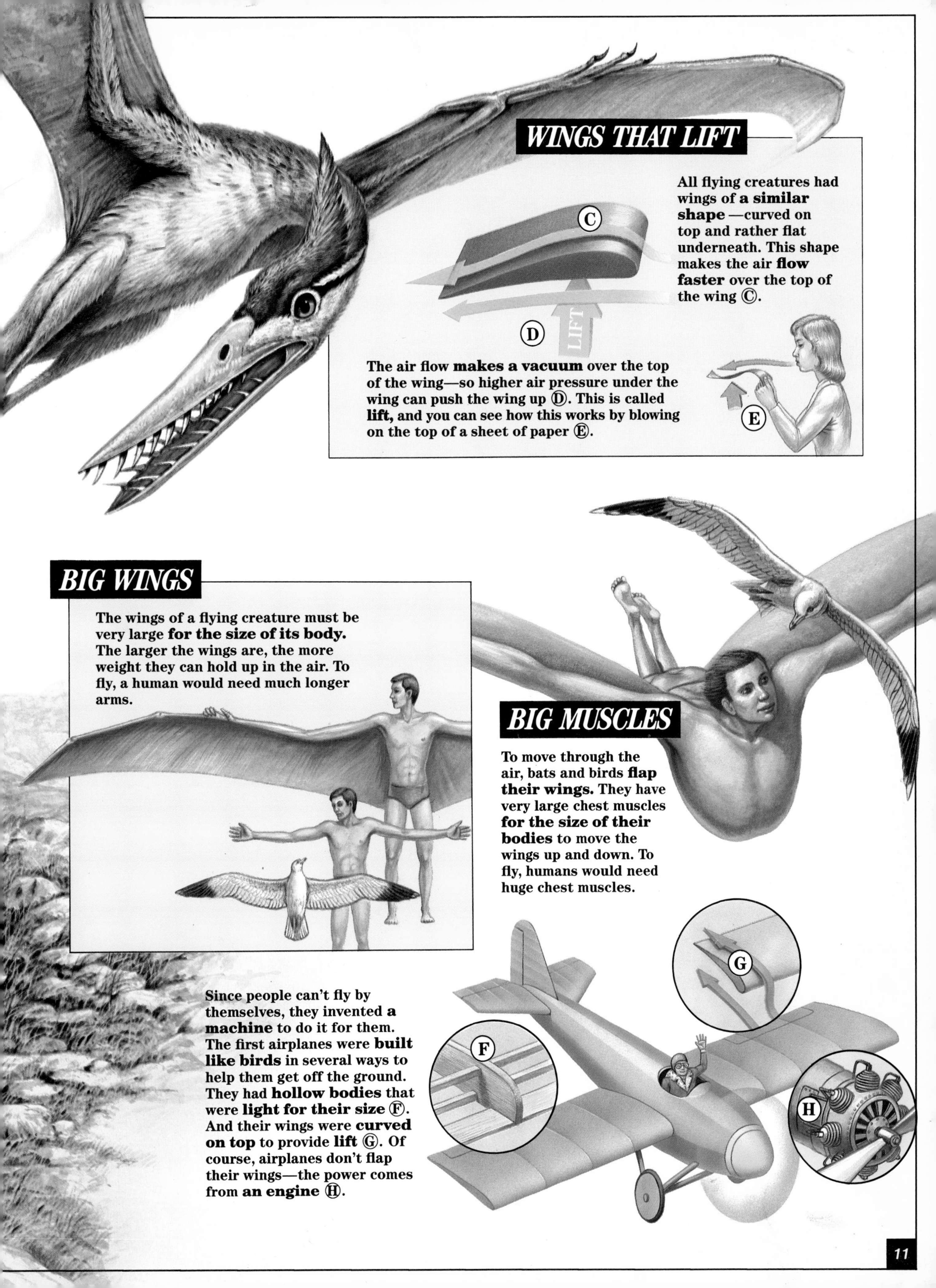

WINGS THAT LIFT

All flying creatures had wings of **a similar shape** —curved on top and rather flat underneath. This shape makes the air **flow faster** over the top of the wing Ⓒ.

The air flow **makes a vacuum** over the top of the wing—so higher air pressure under the wing can push the wing up Ⓓ. This is called **lift,** and you can see how this works by blowing on the top of a sheet of paper Ⓔ.

BIG WINGS

The wings of a flying creature must be very large **for the size of its body.** The larger the wings are, the more weight they can hold up in the air. To fly, a human would need much longer arms.

BIG MUSCLES

To move through the air, bats and birds **flap their wings.** They have very large chest muscles **for the size of their bodies** to move the wings up and down. To fly, humans would need huge chest muscles.

Since people can't fly by themselves, they invented **a machine** to do it for them. The first airplanes were **built like birds** in several ways to help them get off the ground. They had **hollow bodies** that were **light for their size** Ⓕ. And their wings were **curved on top** to provide **lift** Ⓖ. Of course, airplanes don't flap their wings—the power comes from **an engine** Ⓗ.

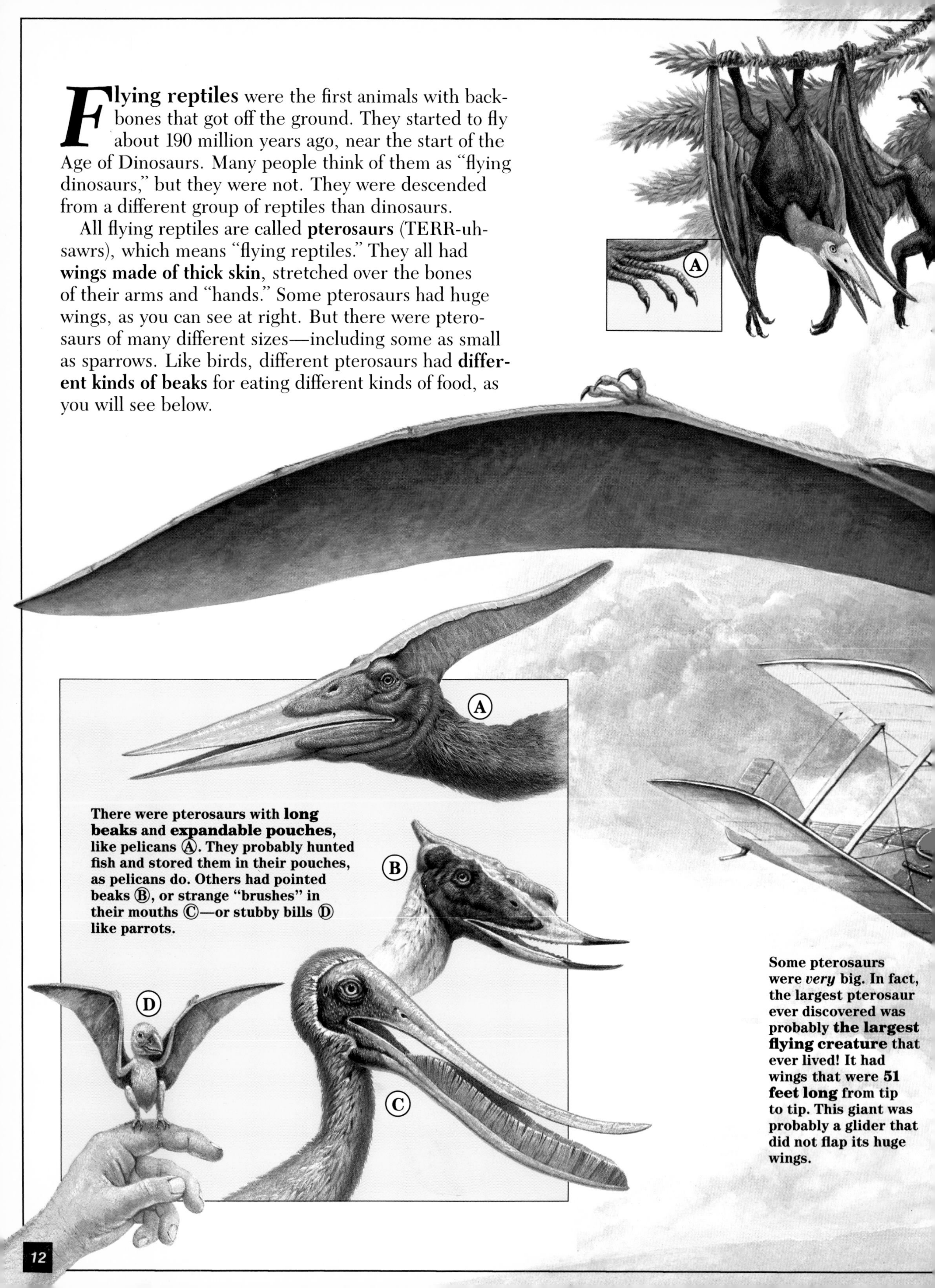

Flying reptiles were the first animals with backbones that got off the ground. They started to fly about 190 million years ago, near the start of the Age of Dinosaurs. Many people think of them as "flying dinosaurs," but they were not. They were descended from a different group of reptiles than dinosaurs.

All flying reptiles are called **pterosaurs** (TERR-uh-sawrs), which means "flying reptiles." They all had **wings made of thick skin**, stretched over the bones of their arms and "hands." Some pterosaurs had huge wings, as you can see at right. But there were pterosaurs of many different sizes—including some as small as sparrows. Like birds, different pterosaurs had **different kinds of beaks** for eating different kinds of food, as you will see below.

There were pterosaurs with long beaks and expandable pouches, like pelicans Ⓐ. They probably hunted fish and stored them in their pouches, as pelicans do. Others had pointed beaks Ⓑ, or strange "brushes" in their mouths Ⓒ—or stubby bills Ⓓ like parrots.

Some pterosaurs were *very* big. In fact, the largest pterosaur ever discovered was probably the largest flying creature that ever lived! It had wings that were 51 feet long from tip to tip. This giant was probably a glider that did not flap its huge wings.

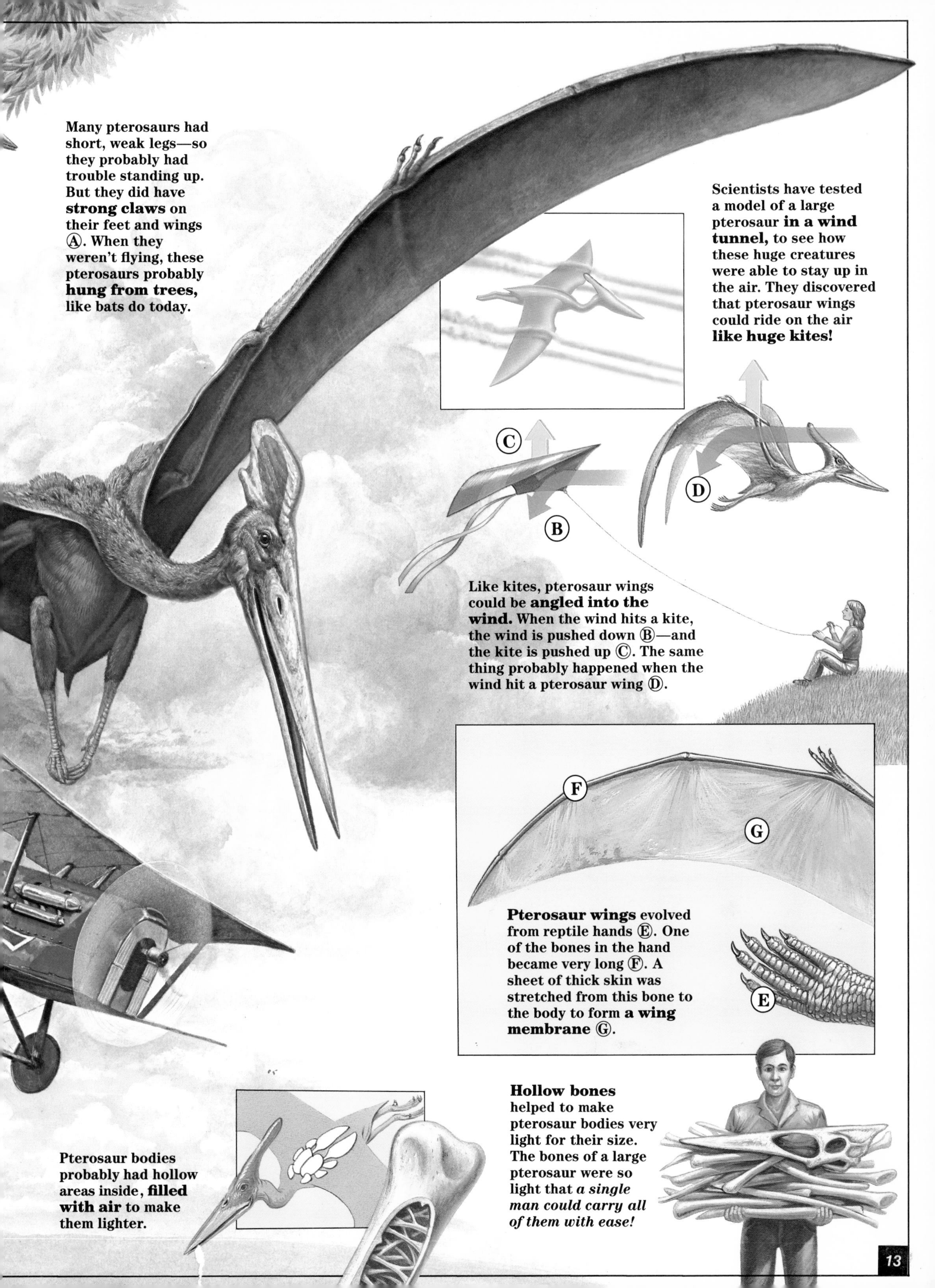

Many pterosaurs had short, weak legs—so they probably had trouble standing up. But they did have **strong claws** on their feet and wings Ⓐ. When they weren't flying, these pterosaurs probably **hung from trees,** like bats do today.

Scientists have tested a model of a large pterosaur **in a wind tunnel,** to see how these huge creatures were able to stay up in the air. They discovered that pterosaur wings could ride on the air **like huge kites!**

Like kites, pterosaur wings could be **angled into the wind.** When the wind hits a kite, the wind is pushed down Ⓑ—and the kite is pushed up Ⓒ. The same thing probably happened when the wind hit a pterosaur wing Ⓓ.

Pterosaur wings evolved from reptile hands Ⓔ. One of the bones in the hand became very long Ⓕ. A sheet of thick skin was stretched from this bone to the body to form **a wing membrane** Ⓖ.

Pterosaur bodies probably had hollow areas inside, **filled with air** to make them lighter.

Hollow bones helped to make pterosaur bodies very light for their size. The bones of a large pterosaur were so light that *a single man could carry all of them with ease!*

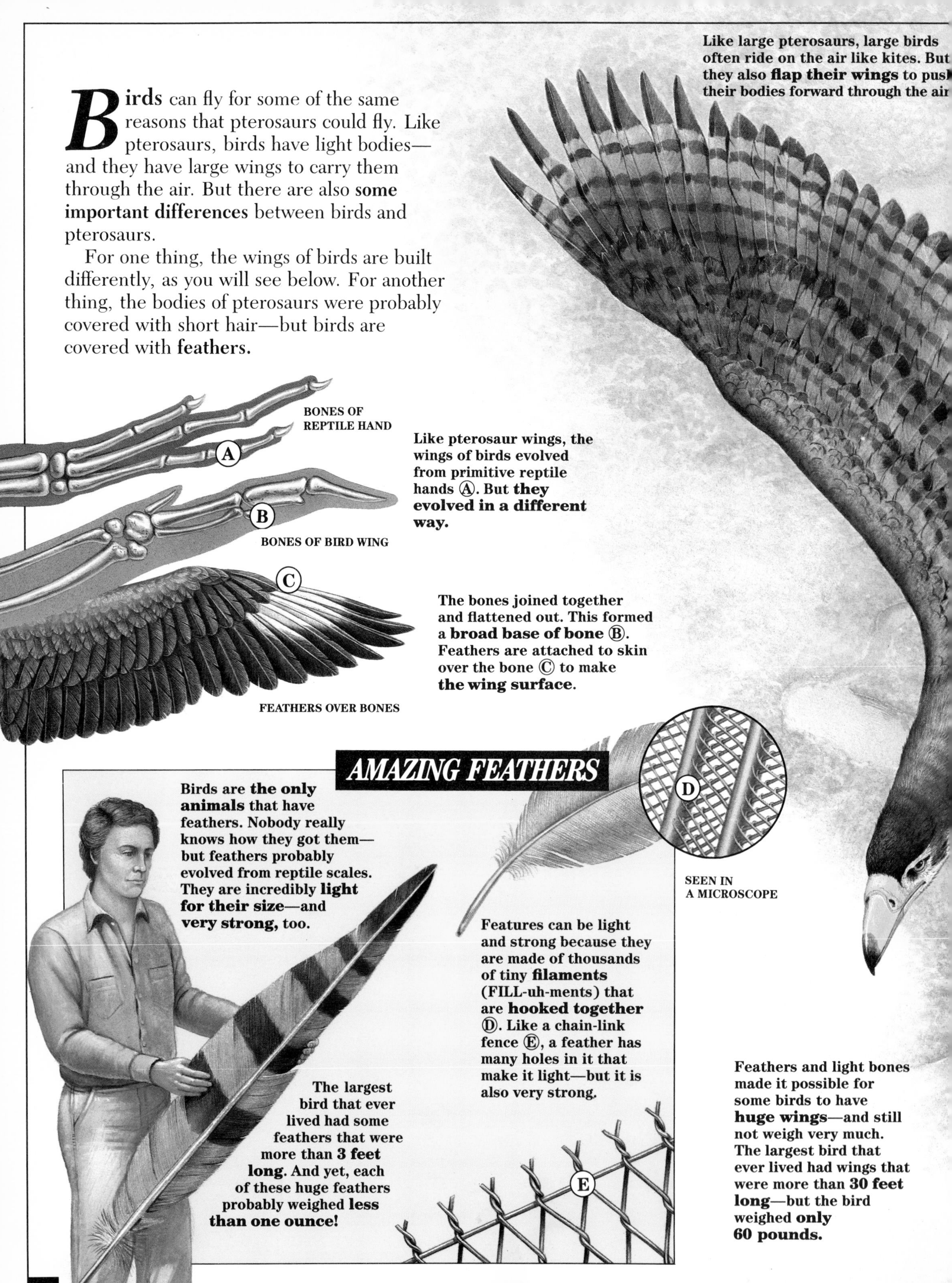

Birds can fly for some of the same reasons that pterosaurs could fly. Like pterosaurs, birds have light bodies—and they have large wings to carry them through the air. But there are also **some important differences** between birds and pterosaurs.

For one thing, the wings of birds are built differently, as you will see below. For another thing, the bodies of pterosaurs were probably covered with short hair—but birds are covered with **feathers.**

Like large pterosaurs, large birds often ride on the air like kites. But they also **flap their wings** to push their bodies forward through the air.

Like pterosaur wings, the wings of birds evolved from primitive reptile hands Ⓐ. But **they evolved in a different way.**

The bones joined together and flattened out. This formed a **broad base of bone** Ⓑ. Feathers are attached to skin over the bone Ⓒ to make **the wing surface**.

AMAZING FEATHERS

Birds are **the only animals** that have feathers. Nobody really knows how they got them—but feathers probably evolved from reptile scales. They are incredibly **light for their size**—and **very strong,** too.

Features can be light and strong because they are made of thousands of tiny **filaments** (FILL-uh-ments) that are **hooked together** Ⓓ. Like a chain-link fence Ⓔ, a feather has many holes in it that make it light—but it is also very strong.

The largest bird that ever lived had some feathers that were more than **3 feet long**. And yet, each of these huge feathers probably weighed **less than one ounce!**

Feathers and light bones made it possible for some birds to have **huge wings**—and still not weigh very much. The largest bird that ever lived had wings that were more than **30 feet long**—but the bird weighed **only 60 pounds.**

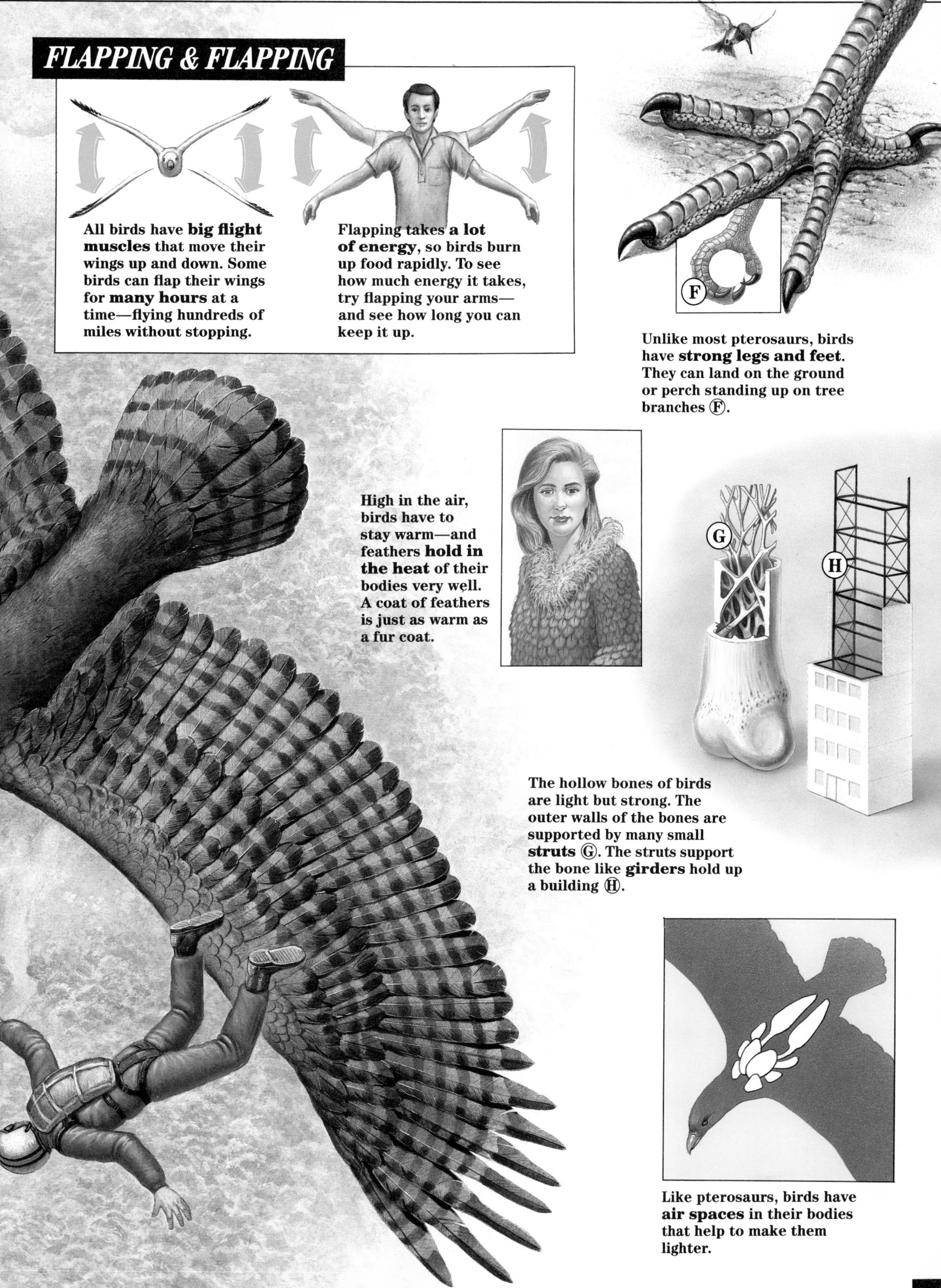

FLAPPING & FLAPPING

All birds have **big flight muscles** that move their wings up and down. Some birds can flap their wings for **many hours** at a time—flying hundreds of miles without stopping.

Flapping takes **a lot of energy**, so birds burn up food rapidly. To see how much energy it takes, try flapping your arms—and see how long you can keep it up.

Unlike most pterosaurs, birds have **strong legs and feet**. They can land on the ground or perch standing up on tree branches Ⓕ.

High in the air, birds have to stay warm—and feathers **hold in the heat** of their bodies very well. A coat of feathers is just as warm as a fur coat.

The hollow bones of birds are light but strong. The outer walls of the bones are supported by many small **struts** Ⓖ. The struts support the bone like **girders** hold up a building Ⓗ.

Like pterosaurs, birds have **air spaces** in their bodies that help to make them lighter.

Where did birds come from? Scientists have several different opinions about this. Some say that birds are **cousins of dinosaurs**—animals that **descended separately** from the same early reptiles that were the ancestors of the dinosaurs. Others say that **birds *are* dinosaurs**, with wings and feathers added.

As you will see, fossils of early birds seem to show that they were **part bird** and **part dinosaur.**

Some small dinosaurs looked very much like **birds without feathers.** Like birds, these dinosaurs had long necks with small heads. They also had **hollow bones.**

A STRANGE FOSSIL

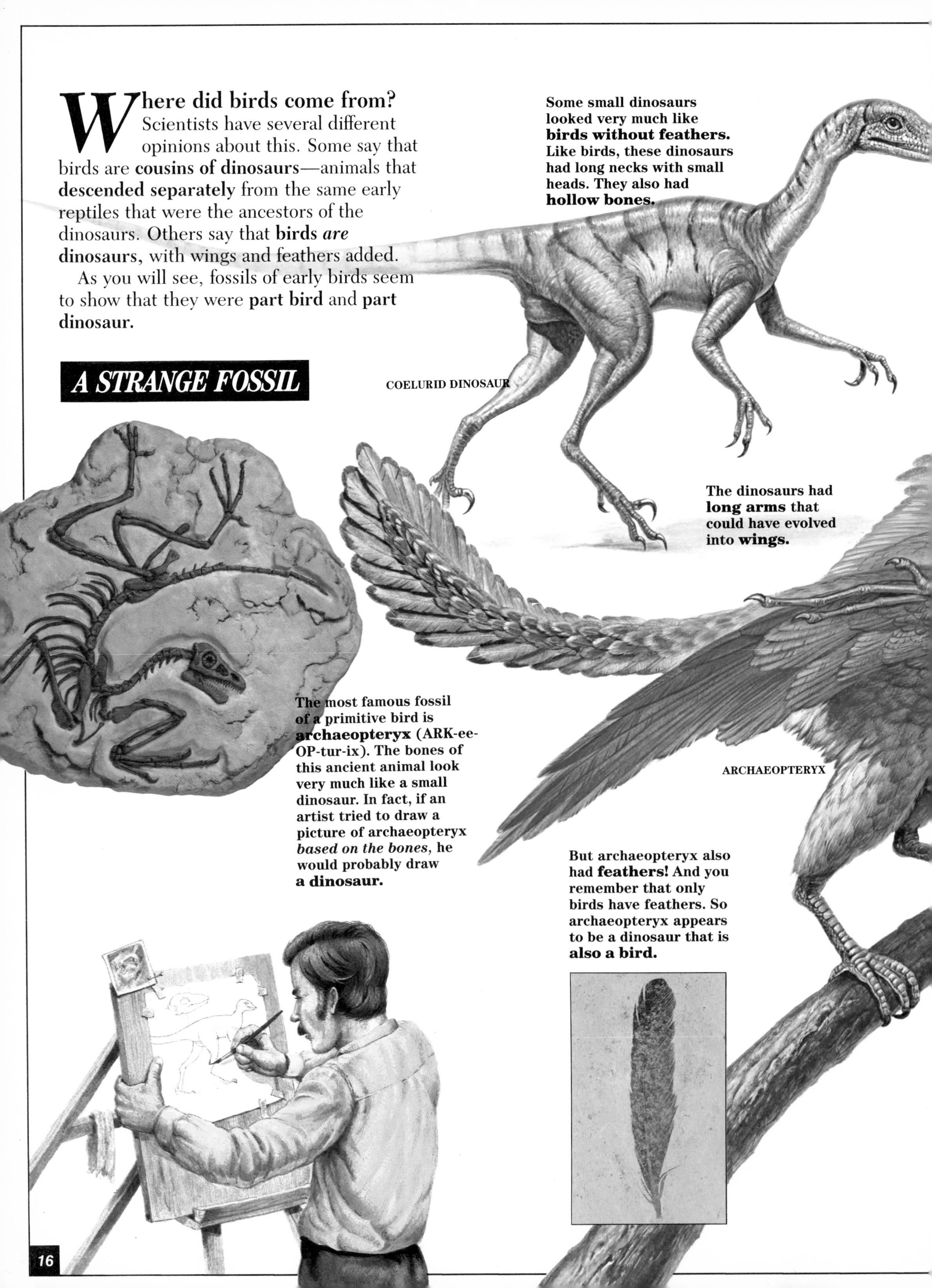

The dinosaurs had **long arms** that could have evolved into **wings.**

The most famous fossil of a primitive bird is **archaeopteryx** (ARK-ee-OP-tur-ix). The bones of this ancient animal look very much like a small dinosaur. In fact, if an artist tried to draw a picture of archaeopteryx *based on the bones*, he would probably draw **a dinosaur.**

But archaeopteryx also had **feathers!** And you remember that only birds have feathers. So archaeopteryx appears to be a dinosaur that is **also a bird.**

HOW DID BIRDS START TO FLY?

There are several theories about this. Some scientists say that bird ancestors were reptiles that went **up in trees** to hunt insects Ⓐ. Sometimes, when insects flew away, **the reptiles jumped** after them Ⓑ.

Over millions of years, the reptiles developed **small wings** to help them **glide** after the insects Ⓒ.

The wings grew larger until the reptiles could **fly**—and they became birds Ⓓ.

Another theory says that small dinosaurs chased insects **on the ground.** They developed small wings as **a kind of net** to help them trap insects Ⓔ.

The wings grew larger—until the dinosaurs could fly Ⓕ. And they became **birds.**

The skull of archaeopteryx is similar to the skull of a bird in many ways. It has the same **ring of small bones** around the eye Ⓖ.

The skeleton of archaeopteryx looks partly like a modern bird—and partly like a dinosaur.

It has **a long tail** and **teeth** like a dinosaur—but **a wishbone** and **wings** like a bird.

The fossil shows that archaeopteryx had **feet and legs** that were very much like the feet and legs of birds today. They had **scales** on them, and the toes could probably be wrapped around branches. So archaeopteryx could probably **perch like a bird.**

Birds that lived like dinosaurs were the largest land animals after the dinosaurs died out. For millions of years, mammals were still small, and **giant flightless birds** took over many of the niches that once belonged to dinosaurs. The largest meat-eating animals and the largest plant-eating animals on land were **birds.**

Some of these birds grew **very large**—up to 12 feet tall. And some weighed *several hundred pounds!* They were too big and too heavy to fly, so they gradually **lost their wings** and developed **enormous legs and feet** to carry them over the ground.

Most of the giant flightless birds became extinct a long time ago. But a few types—like ostriches and emus—are still living.

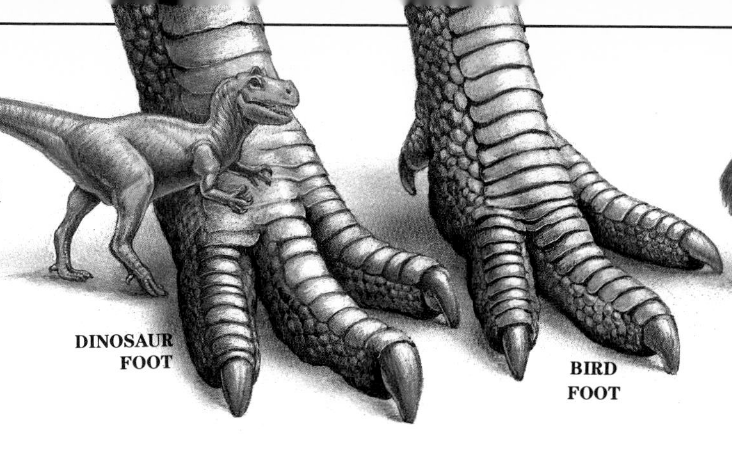

The strong legs and sturdy feet of the giant birds made it possible for them to run very fast, like meat-eating dinosaurs. Their feet even looked like dinosaur feet.

After the dinosaurs were gone, it was safe for birds to come down on the ground to find food. Some birds started to run after prey.

Over millions of years, these birds evolved larger and larger legs, so they could run faster and faster. To catch bigger prey, their bodies got larger.

As their bodies got heavier, they could not fly. They didn't need wings anymore, so the wings got smaller.

To hold up their great weight, flightless birds needed stronger bones. Their bones became more and more solid, like mammal bones.

BONE OF FLYING BIRD

BONE OF FLIGHTLESS BIRD

In a few places, huge flightless birds survived until very recently. On some islands, there were birds over ten feet tall **only a few hundred years ago!**

The birds survived because there were no large meat-eating mammals on the islands to kill them. But then, people came to the islands and killed the huge birds.

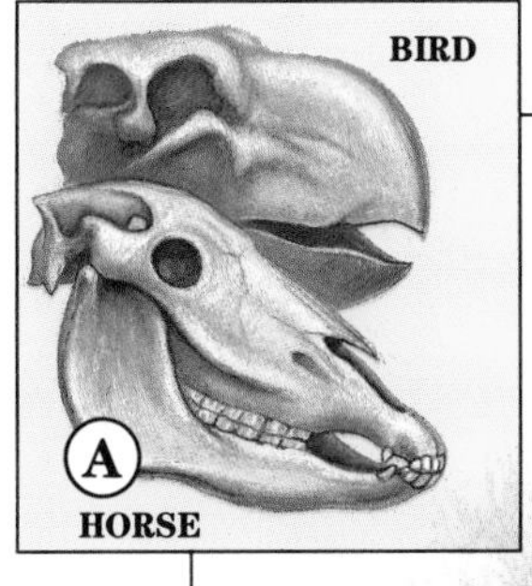

Some flightless birds had **huge beaks** for grabbing prey. The heads of some were as big as the heads of modern horses Ⓐ.

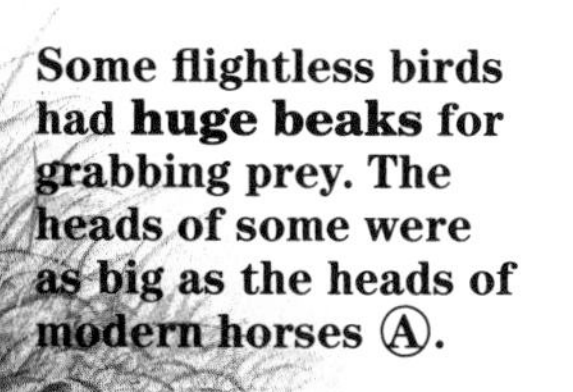

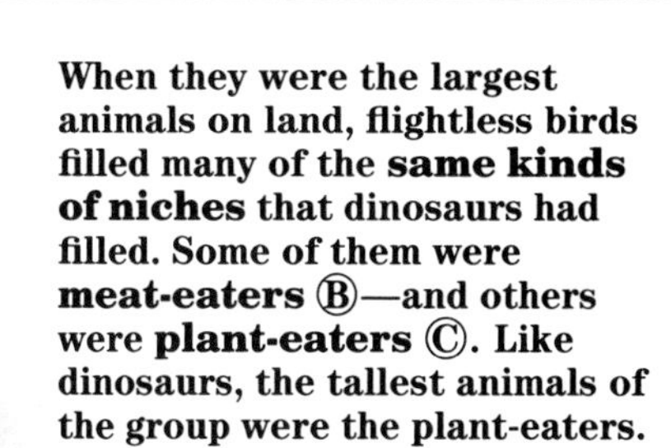

When they were the largest animals on land, flightless birds filled many of the **same kinds of niches** that dinosaurs had filled. Some of them were **meat-eaters** Ⓑ—and others were **plant-eaters** Ⓒ. Like dinosaurs, the tallest animals of the group were the plant-eaters.

The meat-eaters often hunted **small mammals.**

In the end, mammals took most of the niches on land away from the giant birds. Most of the birds **died out**—and mammals became the main land animals.

Bats are the only mammals that fly, and they were the last group of vertebrates that took to the air. Nobody is really sure when they started to fly, because we don't have good fossils of the first bats—but it was probably during the last part of the Age of Dinosaurs.

All bats are descended from small **insect-eating mammals** that lived during the time of the dinosaurs. Like pterosaurs and birds, the first bats may have gone into the air **to chase after flying insects**.

Like birds, bats have been very successful. There are *thousands* of different kinds—and almost *one-quarter* of all mammal species living today are bats.

Like some pterosaurs, bats have weak legs—so they usually hang themselves up when they aren't flying. Unlike pterosaurs, bats always hang upside down.

The wings of bats look like pterosaur wings—membranes of skin stretched out on long bones. But bat wings have four long fingers to support the membrane.

The oldest bat fossil ever found is about 50 million years old. But there were probably bats on earth at least 65 million years ago.

BAT CAVE

Most bats are still insect-eating animals, like their mammal ancestors. They live in huge colonies that sometimes include more than 50 million bats. Like their ancestors, these bats hunt at night. They sleep in caves or other shelters during the day—and leave their shelters at dusk to search for prey.

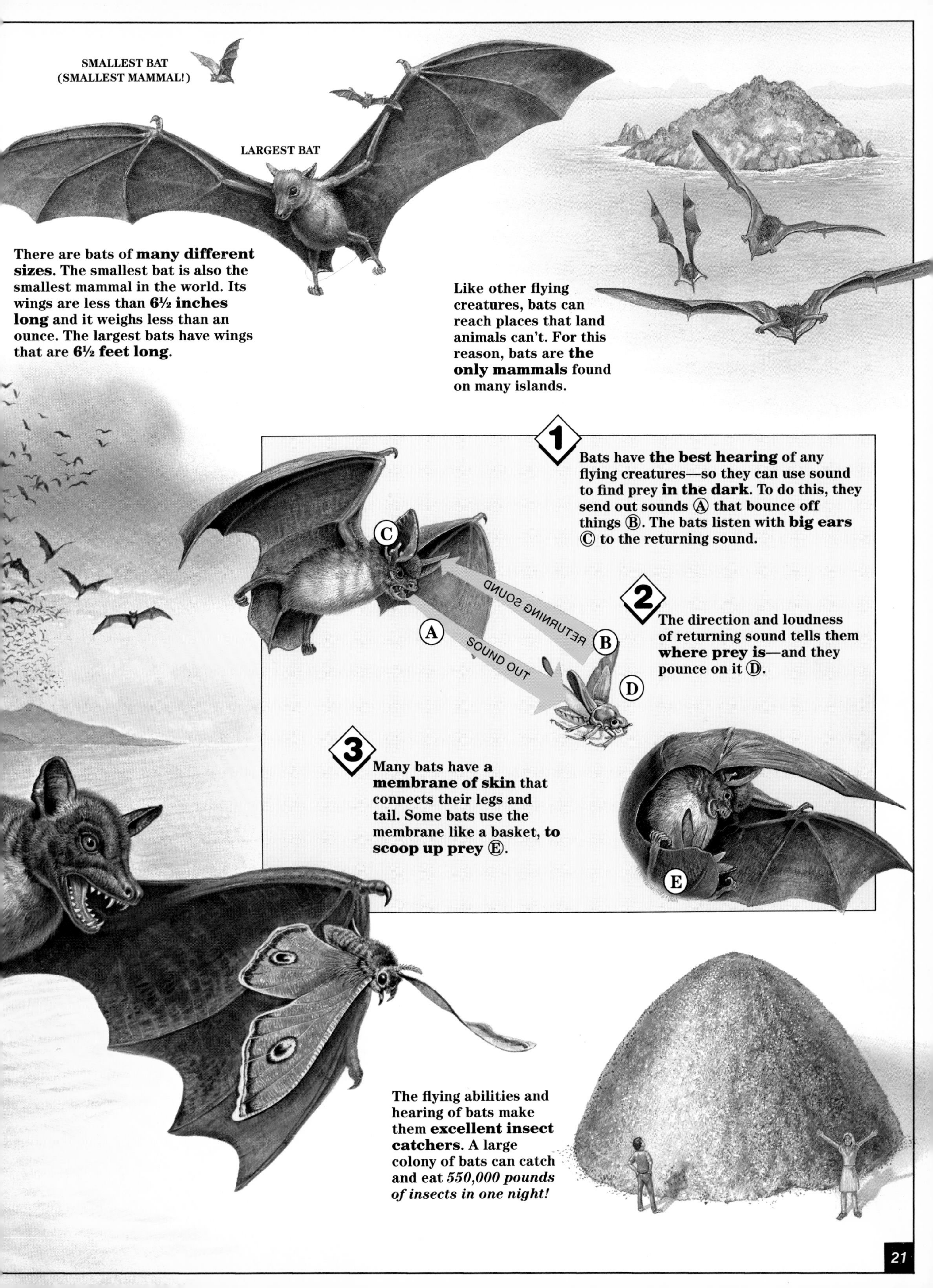

There are bats of **many different sizes.** The smallest bat is also the smallest mammal in the world. Its wings are less than **6½ inches long** and it weighs less than an ounce. The largest bats have wings that are **6½ feet long**.

Like other flying creatures, bats can reach places that land animals can't. For this reason, bats are **the only mammals** found on many islands.

1 Bats have **the best hearing** of any flying creatures—so they can use sound to find prey **in the dark.** To do this, they send out sounds Ⓐ that bounce off things Ⓑ. The bats listen with **big ears** Ⓒ to the returning sound.

2 The direction and loudness of returning sound tells them **where prey is**—and they pounce on it Ⓓ.

3 Many bats have a **membrane of skin** that connects their legs and tail. Some bats use the membrane like a basket, **to scoop up prey** Ⓔ.

The flying abilities and hearing of bats make them **excellent insect catchers.** A large colony of bats can catch and eat *550,000 pounds of insects in one night!*

REMEMBER:

1 Some animals started to fly because there were **great advantages** in flying. They could fly to places that other animals could not reach . . .

2 . . . and they could find **new sources of food.**

3 Animals that could fly were **much safer** than ground animals. For one thing, dinosaurs could not follow them into the air.

4 They could build nests in safer places, **to keep predators away** from their babies.

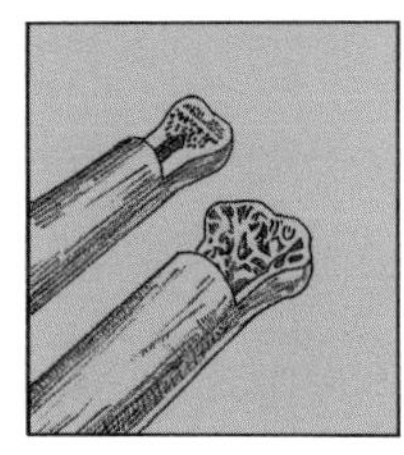

5 It took **special kinds of bodies** to fly. The bodies had to be light, so flying vertebrates evolved **hollow bones.**

6 Flying creatures needed **wings** to hold them up in the air. Some flying reptiles had the largest wings of all—up to 51 feet long!

7 The wings of pterosaurs were made of thick skin called **a wing membrane.** The skin was stretched from **one long finger** to the body, to make a wing surface.

8 Birds had light bodies and wings, too. But the wings of birds were **built in a different way.**

9 Bird wings were covered with **feathers.** The feathers were very light, but very strong. And they could be very big—up to three feet long.

NEW WORDS:

Lift:
The force that pushes a wing up into the air. The shape of a wing causes it to have lift.

Pterosaurs
(TERR-uh-sawrz):
Flying reptiles, probably the first vertebrates to fly.

Wing Membrane:
Thick skin that is stretched to make a wing surface for pterosaurs and bats.

10 Pterosaurs were not closely related to dinosaurs—but **birds might be dinosaurs**, with wings and feathers added.

11 Early birds like archaeopteryx look like small dinosaurs in many ways. They have **the bones** of dinosaurs and **the feathers** of birds.

12 After dinosaurs died out, **giant flightless birds** became the largest animals on land for a time. They took over some of the niches that once belonged to dinosaurs.

13 The last vertebrates to fly were **the bats**. They took to the air near the end of the Age of Dinosaurs.

14 **The wings of bats** are made of thick skin, like the wings of pterosaurs. But the membranes are stretched over **four long fingers**, instead of one.

Filaments
(FILL-uh-ments):
Tiny rods that are hooked together to make feathers that are strong but light in weight.

Struts:
Small rods of bone inside the hollow bones of birds. The struts prop up the walls of the bones to make them stronger.

Archaeopteryx
(ARK-ee-OP-tur-ix):
An early bird that looks like it was part dinosaur and part bird. Seems to show that birds are dinosaurs.

Index